Santiago Roel R

I0476706

Information

The Key to Understanding Complexity

Santiago Roel R

Information: The Key to Understanding Complexity

ISBN-13: 978-1511818902

ISBN-10: 1511818905

Santiago Roel R

Preface- a personal note

In my life I have been confronted with different life views, but basically two competing views that tend to create a conflict between the "spiritual" and the "real", the *other* world and *this* world.

As a young child, before starting school, I would sometimes just know why events had happened, were happening or would happen. Adjusting to life as a new *medium* was not easy. At that time, I asked what I was doing here. What I was really asking was *what am I doing in this dimension*? I soon learned that this whole, instant connection was not common to everyone or not everyone was willing to talk about it, so I kept it to myself. Yet, not having any religious upbringing or mythical explanation I was left on my own to explore.

In my primary school years it was all about learning to be that person I was becoming and to relate to my new environment. I kept trying to resolve the conflict between the social rules and the *other* rules and reach a functioning balance. I would oscillate between being very sociable and then spending a great deal of time by myself.

In my adolescence, the conflict surfaced again and I became withdrawn and philosophical. I took long walks, spent my money on books and learned some yoga. While living in the countryside or travelling alone before college, I experienced strange synchronic events. I finally resolved this phase by becoming social again, finished college, started a business and married. The *real* world pulled me out of my introspection.

At 33 I went through a serious crisis, after which I stopped thinking so much about myself and stepped into government reform. It was very energetic. I also began writing as a twice-a-week editorialist for a newspaper. Many times, an hour before due time, I would just sit in front of my computer with a blank mind and let the article emerge. It would appear effortlessly, I just had to fill in the details. These unconscious, spontaneous articles were usually the most appreciated. I also went back to therapy, which helped me to understand conflicts from childhood and once again, cope with the *real* world. At the same time, I became interested in the more transcendental philosophies and somewhat fluent in archetypes and symbolism. I rediscovered and reconnected to the *magical* side of life. At the end of my therapy, just for the fun of it, I tested the newly acquired knowledge with my psychoanalyst- a highly skeptical person- and gave him an archetypal interpretation. In spite of not knowing anything about him, as is usual in therapy, I described very precise information about his life, his lessons, his character and his dreams. He was, to say the least, shocked.

I was trying to integrate the different sides of my Self. Each new phase of my life would bring an initial liberation, a period of benefits -spiritual, intellectual, emotional and material- and then a new crisis where everything would dissolve. With each crisis, I would disintegrate and reintegrate into a new version of myself but in a different level of consciousness.

Eleven years ago, at the beginning of a new crisis, my parents died within months of each other. I helped them to pass on to the other world in a very mystical experience. This would turn out to be a longer and deeper crisis -as mid-life crises usually are. One of the members of my family became afflicted with what we now know is epilepsy in the temporal lobe. We learned –the hard way- the grave dangers of linear, unwholesome medicine that pushed the family into chaos. Having a better understanding of symbolism, I used meditation and music to reconnect to the Universe. I learned (the whole family did) how to heal myself and to heal others.

In this period, I was drawn back into government consulting, specifically on crime prevention. The crime-prevention model we developed was successful but I didn't know why, so I became interested in theory, mostly on complexity. I encountered this dichotomy between what I knew worked from experience and what authors proposed from a theoretical side.

I do not conform easily to established truth and authority but this time, I am not trying to oppose anything; I am trying to integrate.

Information: The Key to Understanding Complexity

Introduction

In June 2010, I published **How does Order Emerge in Social Systems?** Studying authors on Complexity, I found interesting concepts that helped me to explain our crime-prevention model: sensitivity to initial conditions, the impossibility of predicting precise outcomes, the non-linearity of life, the self-similarity of the system at different scales, how complexity emerges from the iteration of simple rules, the tendency of the system to fall into a set of results called an *attractor*, but most importantly, that order emerges naturally in a system.

Complex social systems cannot be controlled. In fact, the control paradigm is contrary to creating a better environment or obtaining better performance; it is much more powerful to enhance the system's own capacity to self-organize. (I find *self-organize* a much better term than *self-regulate*, which has a mechanistic connotation).

I stated "The control paradigm is highly ineffective and costly because it is unnatural. The worst policies and the most counterproductive managerial, economic, social, educational or political systems come from trying to control other people. Control should be considered an extreme and temporal measure. Rules that work with the autonomy of the parts are a much better choice."

How does that fit into our model?

The crime-prevention model we have developed creates the conditions for order to emerge in a natural way. We do not make cumbersome plans; we focus on results, we observe the system from the "outside" with a holistic perspective; we measure outcomes on a monthly basis and make constant adjustments; we work across government agencies and very closely with the community. As the model iterates month by month, strategies are constantly tested and reinforced or discarded.

From experience and historical graphs, we usually know when a certain crime will peak and prevention actions are directed to counter this expected tendency, but precise prediction is not the goal, the objective is to keep the team aware and creative both for common causes or extraordinary events. There is constant learning and adaptation to the system's performance. This is the basic model:

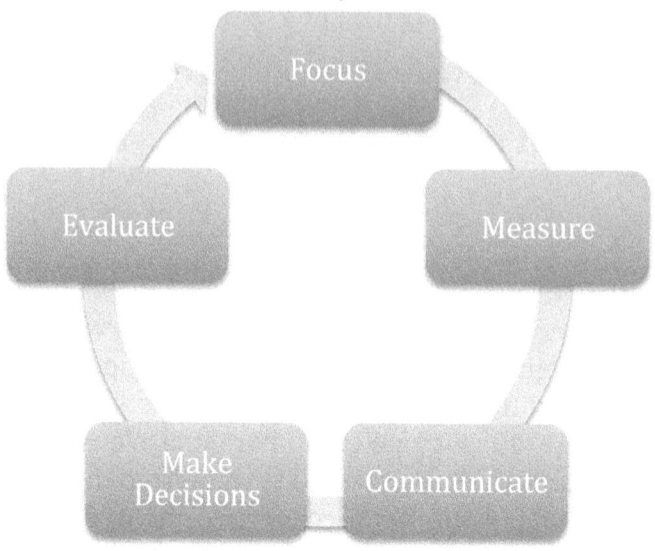

1. **Focus**. This is where we start, where attention and intention are clarified. What will we observe? What do we expect to happen? What is our purpose? What do we want?

2. **Measure**. Rather than just measuring we extract *relevant* information from the system. We try to understand what the system is "saying". We focus on outputs and outcomes of the system. Both quantitative and qualitative information are relevant to understand the system.

3. **Communicate**. We send back relevant information to the system. Information has to be useful for the desired outcome.

4. **Make Decisions**. <u>All</u> the parts of the system make decisions. There is no attempt to control; it is impossible and undesirable to

control the decision process. We do emphasize common goals and teamwork across government agencies and with the community.

5. **Evaluate**. We extract information to evaluate results against the desired outcome.
6. **Iterate**. Repeat the cycle and let the system learn. Information and knowledge are created as the system iterates.

Crime data is a great source of information since it can be obtained as frequently as desired: daily, weekly or monthly. This allows rapid iteration. We also use surveys and knowledge from the people that are close to the customer.

This is quite different from a traditional crime program -or any other conventional government or business program for that matter-which tends to focus on pre-determined actions and a desire to control activity in a hierarchical manner. In a traditional crime program, actions are usually investments in equipment and training –mostly for police- with the purpose of improving *reaction* to emergencies. Traditional programs may also include prosecution enhancement. Both prompt reaction and expedite prosecution are certainly useful but they fall short of radically reducing crime for several reasons:

- They are activity-based and miss the most important point- the outcome.
- They do not work with the *whole* system.
- They are reactive instead of *preventive*.
- Decision-making is slow and budget oriented.

- They do not allow the system to learn from experimentation and innovation.

This linear, hierarchically controlled, pre-determined way of "solving" problems is not limited to crime or government, but applies to most organizations. Instead of allowing complex order to emerge in a natural way, traditional organizations create chaos, which is visible in failure, excessive regulation, fear, bankruptcy, poor customer and employee satisfaction, costly operations, inefficiency, etc. The "traditional" way of doing things becomes an obstacle for understanding integrally and for reinforcing all the system's capacity to constantly adjust to the surrounding environment.

How do we compare to traditional models?

In crime prevention we obtain much better results than traditional programs. All crimes decrease and most of them decrease radically, anywhere from a 25% to a 50% reduction. All of this happens within a period of 6 to 24 months.

In comparison, traditional programs fare very well when they can reduce crime by 5 or 10%. Yet, if a new situation emerges the organization is slow to react and so crime rates can climb back very easily. There is not much learning involved and most of it is concentrated in the hierarchical leader, if the "top" person leaves, things go back to "normality" i.e., constant crisis.

A natural model to learn from

Our model applies not only to crime prevention but also to any kind of social system, including the government, NGOs, communities and business. My "expertise" or experience is not in crime, but in decision-making processes in complex environments.

The model is successful because it seeks to follow the rules of how humans learn by experimentation. As humans, we do this all the time; we have done it since birth. This is the way we learn to walk, talk, ride a bicycle or drive a car. We do it when we mingle freely in a party or walk in the park. We do it when we play music or practice a sport. We do it as kids and as parents. We do it as long as we are not in a traditional organization: in front of a desk in a classroom or at work. We do it when we are most allowed to be what we are: humans.

Take a look at this great example of Sugata Mitra's experiments on self-organizing in learning: http://www.youtube.com/watch?v=dk60sYrU2RU&feature=feedf

We have done it to evolve but somewhere along our history, as we began to live in larger communities and created more complex structures of governance, we designed unnatural ways of solving problems and organizing ourselves. We produced and accepted hierarchical structures to control and dominate others.

We became successful in understanding simple, independent mechanical systems, believed this was how the Universe worked and expanded our intention to create mechanistic organizations. We became obsessed with prediction and fearful of experimentation and thus, created plans with the fantasy that the future could be foreseen and controlled, but control is just an illusion and contrary to learning and evolution. The world is not linear or predictable as classical mechanics assumes.

Not everyone works like this. The highly successful organizations, the ones from which creativity and constant adaptation are demanded by their environment, organize themselves in a more natural way: hierarchies are trimmed or minimized and plans are seen more as a purpose than as a straitjacket. The leaders of these organizations focus on the most important factors: purpose, simple rules of engagement, the flow of information and maintaining a creative environment. Ideas are allowed to emerge and are tested against performance, not with fear but in a playful manner.

Information: The Key to Understanding Complexity

Santiago Roel R

Information as the ordering factor in a complex system

The most interesting element in our model is communication. We constantly extract information from the system such as crime stats and profiles (*when, where, at what time* and in *what manner* are crimes being committed) and communicate them to all the authorities involved, to the media and -with a specific effort- to communities and social groups at risk. The flow of relevant information is the cornerstone of the model and it is a two-way street: we "hear" what the system tells us through the crime stats, and we "talk" to the system in a very proactive manner.

Most of our effort goes into information and communication with the sole purpose of improving what we call the *preventive intelligence* of all agents involved: potential crime victims, police, other government agencies, schools, NGOs, the media, the community at large.

Information is what allows the system to learn as well as to self-organize and evolve on a monthly basis.

The best examples are found in domestic violence and rape. We have reduced both of these crimes by nearly 75% exclusively with information. We have done this by informing potential victims and offenders about the crime profile. None of these crimes are related to police action (rape is committed mostly by friends and family). Neither can be reduced with improved prosecution or prompt reaction, one of the reasons being that many of these crimes go unreported. Still more relevantly, we are focused on preventing these crimes, not reacting to them. Reaction is costly and is very rarely related to prevention, even when highly successful.

A traditional program-for crime or for any other area of government or business- would instead, focus on rules: laws, regulations or procedures that "need" to be changed or added. Many times, most I would say, this intent to add more rules creates a paradoxical outcome: the effort to control creates chaos instead of order precisely because excessive rules deter creativity, experimentation and learning.

In synthesis, an efficient complex system has very few rules and a great deal of information, therefore information becomes fundamental to explain how order emerges in other complex systems like flocks or swarms or crowds.

Starlings

To explain this, I usually begin my lectures with a video of a flock of thousands starlings gracefully flying at sunset. This is one example: http://www.youtube.com/watch?v=eakKfY5aHmY

The dance they perform is hypnotic, extremely well coordinated, and full of strange patterns and surprising moves. Poetically speaking, this could be the dance of emerging order. While the public is watching in amazement I begin with simple questions:

Can you find a leader?

-No leader.

Where is the plan?

-No plan.

Can you predict what the flock will do?

-No

Is there underline order in the flock?

-Yes

What creates order in the flock?

In this last question, the public suggests answers like: instinct, or genetics, or learned behavior, which are definitely elements in the formula but do not explain the *immediate* emerging order. All of these are what I call *rules of action* but do not shed light on the order being accomplished *instantly*; this can only be understood by information: The starlings transmit information with their movements and receive information from partners, obstacles and free space to move about in. So yes, they follow certain rules - as any other complex system does- this is what they are "determined" or "programmed" to do, but what allows complex order in the flock is *information.* If we blindfolded the birds we would disrupt the capacity for emerging order in the flock, unless of course, the birds could compensate with another sense to receive the required information.

When this is understood the public suddenly becomes animated as they realize the obvious: *information as the main factor for the emergence of complex order.* From there on, we extrapolate to other complex systems like traffic, crowd behavior, sports, business, finance, music, education, economics, social movements, politics and yes, crime.

In India, traffic rules are radically different and this is shocking for any newcomer, as has happened to me. Yet, the flow of cars, buses, motorcycles, rickshaws, people and animals is quite surprisingly harmonious in its own way, and there are very few accidents (although a lot of adrenalin); the traffic somehow *weaves* itself in an organic and mysterious manner. Search "traffic in India" in YouTube or take a look at this funny one to understand what I mean.

http://www.youtube.com/watch?v=KZBuDPx9r44&feature=related

The mystery is solved when you think of *information* as the organizing principle, as opposed to our natural tendency to look at the *rules* of action for understanding order.

Can we extrapolate these concepts to any *other* complex system in nature?

I believe we can.

Rules of Action (inter-action)

Rules of action are given within the system but instant information is not given, it emerges, it is constantly being created by the system. Science makes a great effort to understand and explain these rules. Rules of action can be expressed in physical, chemical, biological or social terms. If we are referring to energy, these rules have been studied from different perspectives: mechanical, magnetic, chemical, thermal, radiant, nuclear, elastic, etc. If we are talking about fields, they are expressed as gravity, electromagnetism, etc.

In communication or information theories, rules of action can also refer to information both from the receiver or the transmitter's perspective.

Rules of action are really both rules of action and *inter-action* because they refer both to the individual and more relevantly, to the relationships of the individual with other individuals of the system, and of the system with its environment.

But the fundamental idea here is to understand the difference between a *given* rule of action and *immediate* information; that shift in focus is what helps us to understand the relevance of information in emergence.

The given rules are predetermined and are a limit of the system towards a specific development, but what allows the emergence of complex order is information.

Information as Key element of Emergence- a paradigm shift

This led me to propose *information as the key element in emergence.*

Is this a new concept?

Of course there are mentions and many chapters and discussions dedicated to information by authors on complexity, but there is an important difference between considering information as just another rule or element of complexity and stating that it is the *key* element or principle that allows, conforms, and creates complexity.

This is my point and it is a paradigm shift.

Intention- a controversy

I also proposed a more controversial principle: that *intention* could affect the system as I had seen it happen in my experience. And so, in analyzing social systems, I concluded:

Action follows information, information follows intention.

But that sounds a bit linear and although this does happen at certain moments, the three elements influence each other, so this diagram offers a better way of visualizing it:

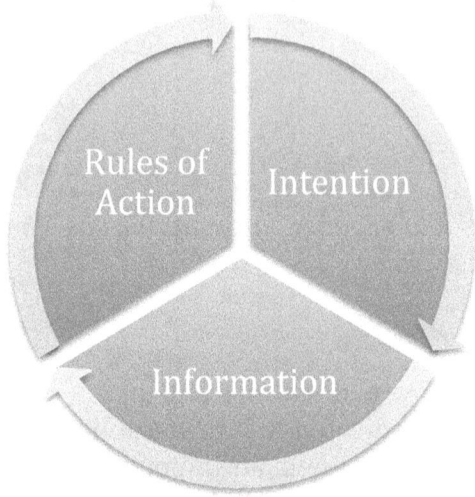

Therefore we have 3 concepts that help us to understand and analyze complexity:

a) The *given rules* of the system.
b) The *novel information* that is created, received, processed or transmitted instantly.
c) *Intention.*

Intention sounds human and therefore, it seems restricted to human systems, but going back to the Starlings video, I do receive some propositions from the audience regarding purpose. Part of the audience is not satisfied with the explanation of rules of action and information, claiming that there is also some kind of *purpose* in the flock, and I agree. Complex systems do have a purpose, something that is projected into the future, a desired outcome.

Mainstream scientists will discard this idea and try to explain higher order solely from the interaction of matter in a *bottom-up* direction, or to express it in my terms, from the exclusive interplay of rules of action and information. Although this might sound "scientifically correct" it becomes much more mysterious and metaphysical than accepting a purpose in the system. In biological systems randomness in itself does not explain complex order and neither does randomness plus basic rules of action. It can create complex patterns that appear to be "alive" because they interact to create something more complex than the parts, but again, they do not explain complex order in a cell or an organism, much less in social or cultural behavior.

An embryo developing into an organism or a seed becoming a tree are extremely mysterious processes because they cannot be explained solely by rules and information. DNA is life's *language* but not life's *program*. We have been led to believe that the genetic code is a program. This is equivalent to saying that letters or words are programmed to become phrases and express meaning or that the random interaction of a binary code (0, 1) will eventually evolve into a computer's software. If that were true, more complex organisms should have more complex genetic codes and this is not so. This has been one of the biggest disappointments for genetics. So the *program* has to be somewhere else.

Information: The Key to Understanding Complexity

Santiago Roel R

Everything *is* information

What I will propose in this essay goes a bit further than solely analyzing the three elements in complexity -rules of action, purpose and information, since all of these three elements are really one: *information.*

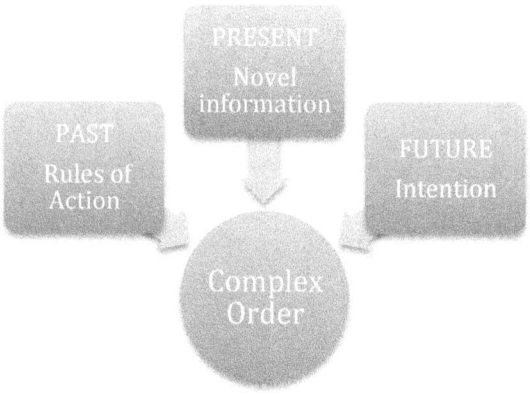

To me, all of the three elements I find for creating complex order are information. It is just a matter of focus.

Rules of action are information that has been formed by previous interactions and iteration; something that the system has *learned.* You may picture this as *crystallized or coherent* information. This is the "stable" part of the system because it has been there for a "long" time and somehow determines the system. Another way of seeing it is like a *canyon* that has been formed by the iterating flow of water, it has become a learned *path*.

Intention is a desired outcome and therefore, it is information that is projected into the future. Where and when that purpose is or was determined is another story. It may be learned from the past or it may emerge as a new intention in the present from novel information, but whatever its origin its direction is toward the future.

Novel information is the information that is presently *emerging* and that *links* the rules and the purpose to create complexity in the present.

Now this, as you may imagine, has many implications and we will explore some of them in this essay. The important part here is to keep in mind the diagram at all times.

Response from the scientific community

Before I advance, let me tell you that the response I have had to these proposals from scholars and the academic community has been mixed: some dismissed it without any thought. Others insisted that at least some of the information proposal is already in the theory of complexity. Most avoided discussing the intention proposal. Those who are more open or less tied to conventional ideas or institutions have agreed, and some have congratulated me on this work.

This feedback has been quite enriching, if not because of its profound counter-arguments, at least because it gives me insight on how the "scientific community" responds to new ideas. I was trying to test my proposals with formal scholars and be open to rejections and suggestions, but these have been scarce, to say the least. So I have moved on to reading as many books and essays on information as I could and trying to test my proposals against the ideas of these authors.

A book that is always useful to keep at hand is Thomas Kuhn's *The Structure of Scientific Revolutions* to understand precisely how "normal science" is based on a set of paradigms. When these given paradigms fail to solve the "puzzles" the paradigms have to be questioned. This opens an arena between competing paradigms, or in terms of Chaos Theory, the system begins to oscillate in search of a higher complex order.

In this process, I now plan to focus exclusively on information and refine my ideas with the intent of trying to define a new useful paradigm from which to theorize, experiment and learn. I will try to tackle this from a fresh perspective.

Caveat

I have to state the obvious: I am no theorist and do not have a "diploma" in complexity. And yet, my lab for the past 20 years has been complex organizations where I have had the opportunity to explore, test, innovate, experiment and learn from experience. All of this, of course, reinforced by the principle that I have had to deliver not a theory for academic purposes but practical solutions for communities and organizations. Customers demand results, and time and resources are always of essence.

I fully embrace the scientific paradigm; I do not feel comfortable with unfounded opinions or un-experienced academic theories or models; I always tend to question the underlying assumptions. I am also aware that science has many unsolved mysteries and in too many cases, this is not sufficiently stated or recognized.

I find "normal science" (to use Kuhn's term) is still influenced by mechanistic or the *matter-is-what-matters* paradigms that become a burden for understanding complexity and life systems; these concepts have yet to be fully questioned by the scientific community.

I value intuition and use it for learning and discovering, but it has to make sense with experimentation and proof. I am very practically oriented and am usually not satisfied until I have tried and experienced first-handedly but of course, not everything can always be experienced and again, not everything can be known.

Finally, concepts are only abstractions, and no matter how we classify them, they are just small pieces of a larger whole.

Where do I stand on information?

I have tried to understand information from the conventional scientific approach that links entropy to information: Sadi Carnot, James Clerk Maxwell, Leo Szilard and Ludwig Boltzmann.

I have also tried to understand the ideas of Alan Turing, Norbert Wiener and Claude Shannon.

I have searched in the bizarre world of quantum mechanics of Max Planck, Niels Bohr, Louis de Broglie, Werner Heisenberg and Erwin Schrödinger, and with special interest, in the interpretation of its cosmological implications by authors like David Bohm.

I have also looked at non-conforming authors who are not afraid to break taboos and propose new ideas with courage and wits such as Rupert Sheldrake, Dean Radin or Ervin Laszlo. As I do not belong to a "scientific community" I can do this freely and without risk, except of course, that of being ignored, which is not a risk if you are already an outsider.

I am not religious in any way. You may recognize an acceptance of metaphysical perspectives, influenced by oriental philosophy, personal experience and authors like David R. Hawkins. Religion is related to the First Chakra – *belonging to a community*; spirituality, on the other hand, is related to the Seventh Chakra, which connects us to the Universe (the larger community). In many ways, science unfortunately, becomes a religion and demands strict compliance to community rules, social classes, existing myths and rituals for membership. In this sense, I am neither religious for religion or for science.

Lastly, and perhaps more relevantly, I have also explored information from my own intuitions, meditations, dialogues and experience. Like Mitra's students, I have never stopped exploring, questioning and trying to create meaning in my life.

When I was about 5 years old I posed my first "profound" question: "Why am I here?" My mother- a psychoanalyst- parked the car, looked at me for a while and gave me the right answer "That is for *you* to find out". So we can read many books and discuss many topics, but in the end, knowledge is personal and it has to make sense for each of us.

In this essay I am trying to understand information in its role with complexity. Some statements may sound laconic or even cryptic. I apologize for that but if I elaborate, they lose the intended meaning or maybe I just don't know more or I can't explain more precisely.

Some proposals are specifically directed to answer or resolve an existing tension between seemingly un-compatible concepts. These are some of my findings:

An inter-connected Universe- a "new" paradigm

The Universe is wholly interconnected; it is not an aggregate of separate and unrelated parts. It connects both in a horizontal (between the parts of the system) and vertical manner (*higher* and *lower* systems). It is connected in space and time, in *potentiality* -what may happen- and *manifestation* - what has happened or is happening. It is a *flow*, it is not fixed or static, it is always *becoming*.

Spiritual mystics would clarify my proposal: for them, the Universe is *One* and *Everything*, there are no separate parts and therefore they are not connected, they are the same. Separateness is *maya,* an illusion of the manifest. Scientists, on the other hand, unless related to the cosmological interpretations of quantum mechanics, would not be interested in this concept. Traditional science focuses on parts not wholes. Finally, religion feeds on dualism and antagonism, so the connection becomes more of a struggle between opposites than a communion of the Whole.

Why the cosmos is connected and why it is so *perfectly* connected and coherent scientists and philosophers can only speculate. That is unknown.

But what we do know is that it is connected in a beautiful and intricate way that allows us to be here and wonder.

This is not new, it has been proposed by many philosophies and spiritual doctrines. It is only new to modern science, which has attempted to explain the Universe with the tools at hand. And so, the unexplainable, that which could not be modeled or spelled out in mathematical terms has been sent into the metaphysical, or worse, it has been discarded completely from scientific attention.

Biological systems and social systems are much more connected and so, are much more complex than physical systems. But this should not deter us from trying to understand them. The less complex realm of physics offers concepts, rules and models that are useful but also limited for understanding the more complex systems of the world we live in. Hence, not because we are scientifically unable to elucidate biological or social complexity from an atomistic view should we embrace the paradigm of a separate Universe.

Complexity returns us to an inter-connected Universe even if we cannot model it, symbolize it or predict its behavior in a precise way. A connected and complex Universe makes much more sense because it matches our own experience and intuitions. Additionally, it is much more helpful to understand the limits of our knowledge than to pretend we know. It is more useful to maintain the tension of a "new" paradigm or between competing paradigms or cosmologies even if this does not solve all the puzzles, than to resort to the comfort of limited science or religious myth, or the other way around, limited religion and scientific myth.

Finally, following Ludwig von Bertalanffy's proposal on systems, complexity also allows us to de-specialize knowledge. We can use concepts across disciplines and explore, much to the dismay of some specialists who don't like opinions if you don't belong to the "club".

So, if we accept the Universe as complexly connected, what connects it?

Information: The Key to Understanding Complexity

Information-my conclusion

Information is what connects the Universe. It is everywhere. It connects the cosmos both in space and in time.

It is a property that is related to energy and matter but which cannot be fully explained from a strict energy-matter perspective.

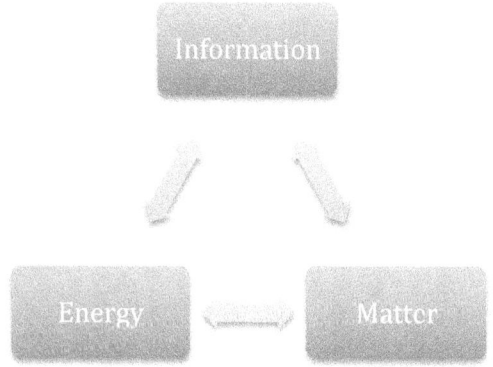

Information can be expressed though energy or matter, information can be transmitted as a wave or as a particle, but is different and independent from both and therefore, information is not necessarily restricted by the limitations of matter and energy.

This is perhaps why there are "spooky" actions at a distance (non-local causality so called by Einstein since they happen "faster" than the speed of light), not requiring time or space for transmission. This means information does not always require mass or energy or a medium to be transmitted or accessed.

Vertical Causality

Another way of seeing it is that information is connected in a vertical way, not in a horizontal space-time manner and therefore creates *vertical causality*. This could explain the *non-local, non-continuous, non-causal* principles of quantum physics.

Horizontal causality is the one we are used to in classic mechanics and the space-time dimension. Vertical causality adds the dimension of information and can explain the sudden appearance of particles from nowhere, complex synchronicity, and unexplainable causality.

Information: The Key to Understanding Complexity

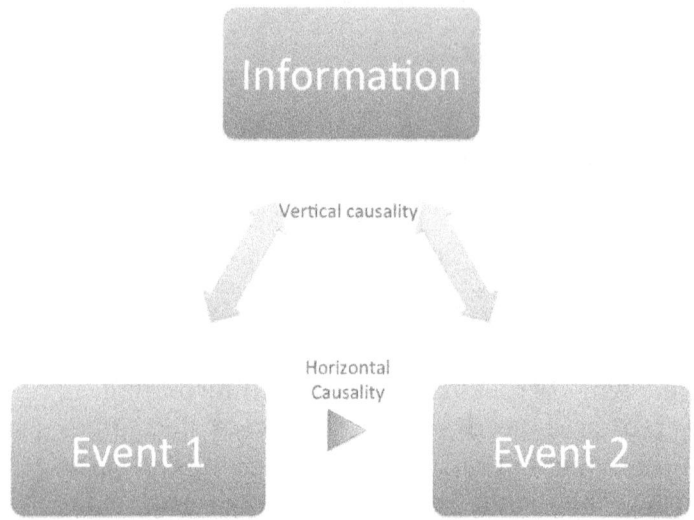

Information as a Field

Information can also be understood as a field, something that influences energy and matter and can only be observed, perceived or measured by its effects like a magnetic or a gravitational field. Something like this.

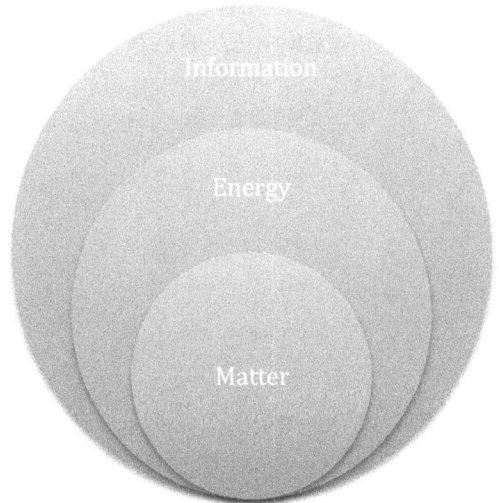

In-formation

Information is created by activity. Everywhere, wherever there is a particle –subatomic or galactic– there are movement and energy and relationships between them. Information is created by these relationships and inversely: information determines new activity, new relationships and new phenomena; this is what is understood from the concept of *in-formation*: The creation of form, movement and mass from information.

Information in short, entangles the Universe in a coherent way. Information relates, connects, creates and in-forms the Universe. It is an organizational element; it allows complex order.

So why have we missed this part? Are we so immersed in information that we don't see it anymore? Are we like a fish that is not aware of the water until it is caught? Is it so obvious we have overlooked it? Is it so intangible that we can't see it from our engrained paradigms? Have we over-focused on matter or energy?

Holarchy and Hierarchy

Holarchy (Arthur Koestler/Ken Wilbur) is a much better concept for understanding information than hierarchy since we tend to equate hierarchy with top-down control. Holarchy on the other hand, enhances levels of integration and order.

Hierarchy

Ecosystem

Organism

Organ

Cell

Molecule

"Holons" are whole and yet, they also belong to a larger set in a "nested hierarchy". An atom is a whole in itself but it can also form part of a larger whole of a molecule. A molecule is a whole in itself but it can also form part of a cell. Is there a limit?

Holarchies are better understood without any limits, so you can go as deep or high as you want.

Holarchy

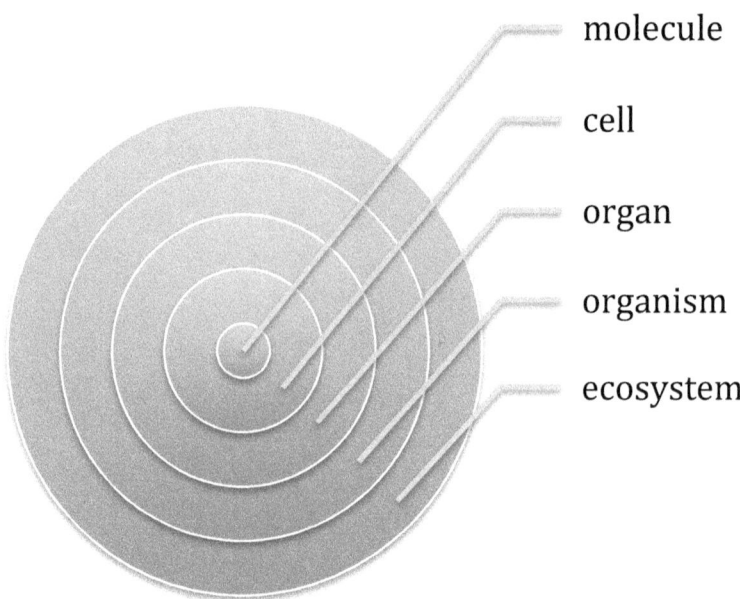

molecule

cell

organ

organism

ecosystem

In this way, we find the three elements of our conceptual model-rules of action, information and intention- in each level: an organism has its own rules of action, processes novel information and has a purpose; but this same organism has organs and belongs to an ecosystem, or a family, a community, a nation, an so on. At every level there are specific rules of action, novel information and intention. And everything is linked and flows.

In our experience, the most successful deployments of our model are accomplished when we are successful at reducing or eliminating hierarchies and enhancing the wholeness of the system.

Information: The Key to Understanding Complexity

How is information created?

Information is created by the interaction between the parts, by the interaction of the parts with the system, by the interaction of the system with its parts and by the interaction of the system with other systems or the environment.

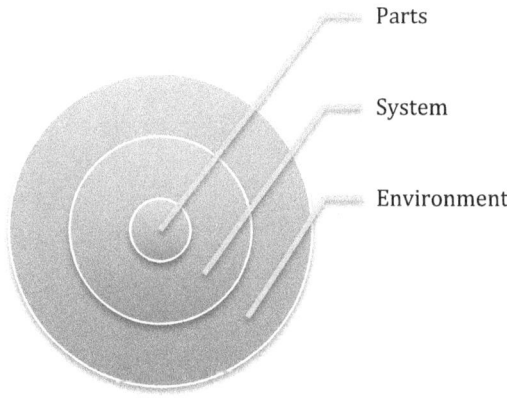

So it is all about relations and inter-connections, even between these two concepts:

Action creates information and information creates action.

Information is also created by the relationship of information with itself.

Bottom-up and top-down emergence

"Top" = the system, the whole, a nested hierarchy of holons, a set.

"Bottom" = the parts, elements of the system, a sub-set.

Information can flow in a *bottom-up* direction or in a *top-down* direction.

Bottom-up information emerges from the interaction and entanglement of the parts.

Top-down information emerges from the interaction or entanglement of the *system* with its *parts*. It can also be understood as purposeful and focused in-formation. This implies that the system has a purpose, which is ampler or larger than the purpose of its parts.

Top-down controversy

Top-down emergence is controversial; it is repealed, avoided or denied by many mainstream theorists since this could lead to a "creationist" point of view, a "God" that creates or controls the Universe. Or maybe they just feel more comfortable and safe with the traditional materialistic, bottom-up, random-god paradigm.

The *cellular-automata theory* (how simple rules can create complexity as they iterate) tries to prove that order emerges in a bottom-up direction without the need for a top-down purpose. The experiments are impressive, the parts seem to come "alive" and generate something much more complex just by interacting within the boundaries of the pre-determined rules.

But can bottom-up emergence solely explain complex order in biology or in society? Can it explain life? Can this fully explain the development of an embryo or social order?

More specifically, is the purpose, in reality, implicitly hidden in the rules established by the designer of the experiment or the program? How is the programmer affecting the outcome? If we program to simulate a flock's behavior-as has been done- are we not influencing activity with an implicit purpose, for example programming it to stay together?

Intention and purpose are so clear when we look at biology or society that we have to break away from this outdated controversy and move on.

Moreover, as you will see further on, intention or purpose could have emerged in a bottom-up direction from previous iterations. I am not proposing –because I don't know- that purpose has been previously created in a Platonic or God-like fixed sense, I feel more comfortable with emerging or iterated purposes, but the fact is, that both rules of action and purpose are present in the system's present behavior.

Nevertheless, science becomes very mysterious, as mysterious or metaphysical as a "God", when it tries to explain complexity or evolution strictly from the random interaction of matter. It is *statistically* impossible to explain the evolution of the Universe from a random bottom-up perspective.

From experience, I know that intention plays a crucial role in emerging social complex order. When our model does not deliver the expected results it always- I underscore- it always has to do with intention. If we fix intention everything aligns properly. So why entangle ourselves with outdated controversies? Why can't we accept purpose, in the top-down or higher-to-lower order, without having to distract ourselves with religion?

Is all information being created instantly or is there given information within the system?

There can also be *given* information in a system. In fact, *rules of action* are really information in a "crystallized" or coherent state. Information that has repeated itself for enough time to become a learned path, an *attractor* (Chaos Theory), a memory, an *archetype* (Jung), a valley, a law, a custom, a program, an instinct, a *morphic* field (Sheldrake), a habit, a genetic code, a *paradigm* (Kuhn), a hologram, a culture, knowledge or a principle.

The Universe "learns" form re-iteration, there is evolution at every level: physical, chemical, biological or cultural.

The existing paradigm in science and religion is contrary to this idea: for both, there are universal, unchangeable laws that have been with us since the "dawn" of time or maybe just the Big Bang, or the previous Bigger Bang. Evolution in science is acknowledged in matter but not in the laws of physics, chemistry or biology that affect matter; the Universe evolves according to a set of pre-fixed rules. For religion these rules are set by a divinity, for science, these laws are simply set, and who or what set them is a metaphysical question that is of no interest to scientists.

The long iterations of the Universe are an unsolved mystery. What was there before *our* Universe? How is it that the Universe has become so finely tuned? Has the speed of light, or gravity, for example, always been the same?

In biology or sociology new rules emerge and evolve constantly. The time scale of these iterations allows us to understand that laws are neither universal nor eternal. Why would this only be true for a part of the Universe? Why do we project our desire to control to the Universe itself?

Information transmission

Information can be created, accumulated, retrieved, related, processed, stored and transmitted in all its states: as mass, energy or purely informational.

At the energy-mass level information is transmitted by particles, waves or fields, it requires some sort of energy for transmission. In the purely informational state, since information does not involve mass, it does not require energy to be transmitted, processed, retrieved or accumulated. Again, once we liberate ourselves from the material paradigm we do not need energy to explain information and it surprises me that this is has not been proposed.

Information: The Key to Understanding Complexity

The I-field

Information is everywhere. It also exists in the "vacuum", the "plenum", the zero-point field, or the Akashic or Informational Field-as Ervin Laszlo proposes. In this state, information is a potentiality, independent of matter, energy, space and time, but related to or entangled to the manifest Universe.

As we move away from matter, there is energy, as we move away from energy there is only information; it is not about "strings" which still sound very material, but about pure information. Are we stuck in science because we are trying to restrict and conform information to matter or energy when it should be the other way around? Are we trying to eliminate dimensions in an attempt to reduce and comprehend? I believe so.

We have explored the physical and have only begun to decipher the strange quantum level, even farther or deeper is the informational dimension.

Other informational fields

There are manifest, specific I-fields where information has been reiterated enough to become redundant, entangled and coherent. It creates a probability of outcome in the system. Both top-down and bottom-up emergence creates these informational fields. This is what Rupert Sheldrake calls *morphic* fields and what Carl G. Jung defines as *archetypes*; informational fields that influence our biological, psychological and social development.

In our crime-prevention model, are we really creating an informational field with our intention and informational effort? It is hard to prove, but there does seem to be a *shift of consciousness* in the system once it reaches a turning point, once it has iterated long enough to create a new order.

Synchronicity and diachronicity

Synchronicity is the simultaneous manifestation of entangled information. Simple synchronicity happens in physical systems (wall-clocks that beat in sync) or in social systems (people unconsciously walking or clapping in sync), but there is also a more complex synchronicity that is not as easy to observe and understand. An example of this mysterious type of synchronicity is that of archetypes where the abstract and the specific meet to express the hidden archetype, the implicit order. But nature is full of them, what can be more mysterious than a living organism or an ecosystem where everything is synchronized in a harmonious way?

Complex synchronicity is hard to understand because causality does not happen in a linear and horizontal way. It does not happen at the same holarchical level, it is non-local, but is related to a higher implicit order that affects the system. It is as if rules of action, purpose and information suddenly appear from another, unseen dimension and affect the system. Many of the present mysteries are related to this and remain a mystery from the stubbornness of science of sticking to the horizontal-linear paradigm.

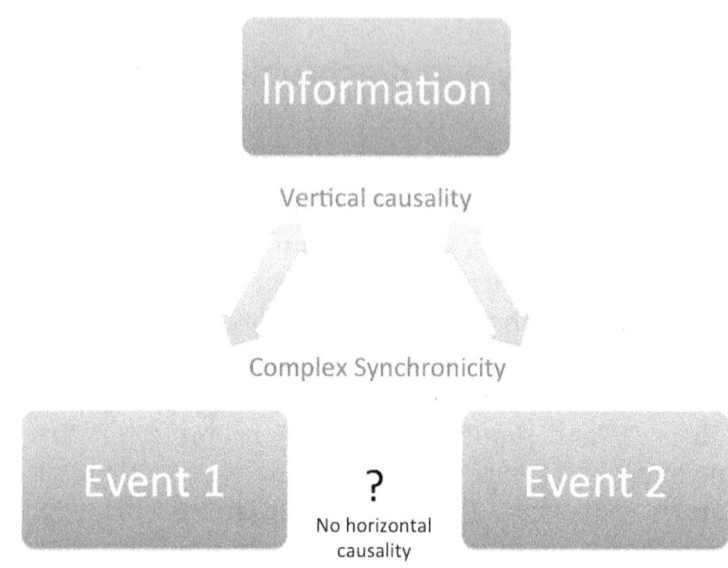

Diachronicity, on the other hand, is the evolution of entanglements through iterations. It is the formation, or in-formation, of synchronicity. It is the process that leads to synchronicity.

In my experience I do observe a strange synchronicity when the system has iterated long enough to influence its environment and its parts, as if we had created an informational field that influences the *whole* system. When we are focused on solving a problem suddenly an event emerges and gives us a clue. Now, the interesting part comes when this clue appears in a symbolic way. Not everyone can read or understand symbolic language but if you are sufficiently "tuned in" an open to these kinds of events, they do happen, and they happen more frequently than we tend to acknowledge; staying focused and creative is the best way to tune in.

Information: The Key to Understanding Complexity

Knowledge

Knowledge is the repetition of entanglements that become coherent. The cosmos "learns" from iteration, which creates paths, rules, laws and coherence.

Knowledge evolves. The system constantly experiments and learns; this is how the Universe develops, expands, grows, and creates laws and principles.

We learn through experience, but can we also access un-experienced knowledge if we re-connect, tune-in, or resonate with the I-field or the "source"? This is what many mystics propose and what the systemic psychology called *family constellations* developed by Bert Hellinger, achieve in a very visible manifestation, and yet, mysterious causal way. If you have participated in a family constellation you know exactly what I mean, if you haven't, it will be hard to understand from a theoretical perspective but you can try. The theoretical explanation for family constellations is based on Sheldrake's concept of morphic fields or information fields.

And so, we can tune-in or entangle ourselves to physical, energetic or informational fields and experience and learn from this connection. This is how our body "learned" to breath air, or how we learned to swim, this is how we learned to cope with gravity and family or social rules. We constantly learn from these experiences.

Focused and random learning

In learning, there are both random and focused activities, both of which are important for learning. The system learns by intentional direction, but also by unintended, accidental or random exploration or reception of information.

In Mitra's experiment with self-organized learning, the kids follow a purpose; Mitra fixes the intention or the purpose in an explicit or implicit way, either by explaining the objective or by simply installing a computer in the wall of a poor neighborhood, and then he allows for random behavior to make its way by giving the students the liberty to explore and learn.

Biological communities have been observed to use both processes for learning. There might be a common purpose of a flock or a community and this creates a focused process of the system. But there is also random, non-focused activity going on that contributes to knowledge and attaining the purpose.

In our crime-prevention model, we always begin by defining the purpose, but from there on, we let the system iterate and the parts learn from random interactions.

This implies that we have to allow enough liberty for a system to learn. We may set the intention, vision or purpose, we may define the rules of engagement, but once that is done, the system has to be left to its own tinkering and exploring in order for information to flow and interact.

If there is enough liberty in the system, new intentions or goals emerge in the process.

This is, I believe, how nature learns.

Attention

Attention is an *intended* action of awareness or observation focused on receiving information or deciphering meaning. Attention influences the performance of the system.

Dean Radin, from the Institute of Noetic Sciences, has documented how human attention focuses on important world events in an unconscious and mysteriously connected manner. The strangeness of it all is that these attention phenomena seem to happen *before* the event. This can be interpreted in different ways: There could be unintended attention or there can be underlying links of intention.

From quantum mechanics we have learned that attention from the observer affects the outcome of the experiment, but I can comment from my own experience in a more mundane way. In our model, what we focus on is extremely important, sometimes it doesn't even make that much sense from an orthodox or logical perspective, and yet, once we have "entangled" ourselves to that specific focus we do influence the outcome. Therefore, what we ask, what we look at, where we focus is of supreme relevance. A paradigm shift in science, for example, is always related to a new perspective of attention, to a new view of the cosmos, to a new window we open to observe the world from our limited human time-space house.

Redundancy

Classical redundancy is usually defined as repetition of information within a message. The more redundancy the less information contained in a message. The less redundancy the more information contained in the message.

Codification of messages or compression of digital archives is accomplished by eliminating the "extra" or redundant information.

THS S N XMPLE F RDNDNCY LMNTN

I eliminated the vowels and you can still understand the meaning.

The genetic code seems to have many redundant series and geneticists don't understand their purpose or function.

I would like to add the concept of redundancy as the *iteration* of the *same* message through time. Redundancy through *iteration* is what creates rules of action. Redundancy, in this sense, is recurrent information that reinforces or ensures the message or the function and/or the probability of the output.

Stable systems produce redundant information. Random systems on the other hand, produce little or no redundancy. Turbulence is full of novel information. Biological systems are somewhere in between extreme order and extreme chaos, they "live" on the edge of criticality where information is neither completely random nor completely redundant and where there is predictability within statistical boundaries or complex attractors. Another way of seeing it is as a mixture of random and redundant information that interplay to produce complex order, or as we have stated, as the interplay of rules of action and novel information.

Redundancy has a purpose; it can be a fail-safe mechanism, it can be the process for creating rules through iteration or it may be a way of ensuring outcomes.

In our model, we do use redundancy both while iterating and in "talking" to the system; we do want to make sure the relevant information reaches everyone and for that we have to be redundant. We have to "move" the system to *iterate redundantly* in order to produce results.

Symbols and archetypes

A symbol represents much more information than expressed.

Information can be compressed, contained or expressed in symbols. Mathematics, and language are good examples of symbolism, but symbols are everywhere.

Symbols can link apparently unrelated information as in archetypes and so become a gateway to knowledge or new meanings. An archetype unfolds, connects and relates. The archetype can express both the concrete and the abstract, and the manifest with the potentiality. An archetype is an attractor but much more complex than an attractor as defined by non-linear dynamics since it is symbolic; archetypes unfold and express all the potentiality of the symbol.

Are symbols and archetypes only related to interpretation by human intelligence or are they everywhere in nature and the cosmos?

Information: The Key to Understanding Complexity

Meaning and relevance

From the receiver's perspective, meaning is the identification, recognition or relation of novel information- a message- with a set of previous entanglements or knowledge. Therefore meaning is always related to context and previous iterations.

But meaning can also be related to a purpose. Consequently, novel information may be interpreted as relevant when it interacts with purpose and/or experience.

The clearer the intention and the more liberty the system is allowed for free iterations the more meaning created by the system through the interaction of its parts.

Information: The Key to Understanding Complexity

Evolution

Evolution cannot be statistically explained only from a bottom-up, random emergence. It includes both bottom-up and a top-down emergence. The system in-forms the parts and the parts in-form the system. Much more relevantly, the system is *in-formed* by its environment.

Evolution-as Sheldrake proposes- is everywhere and not limited to the conventional biological meaning. Evolution is the unfolding path of the Universe in its physical, chemical, biological and cultural manifestations. There are no universal truths or laws; the only universal law it seems, is evolution itself.

Information: The Key to Understanding Complexity

Telos or Omega

Purpose is a highly evaded concept in science. And yet there is purpose everywhere we look in nature. To understand purpose we have to look at function and performance.

There is performance and purpose in parts and in systems. A system's performance is ampler than the performance of its individual parts, it includes the parts but it performs a more complex function.

"Higher" than the system's purpose is the purpose of the environment. Earth –if we observe its performance-has a purpose: Life.

The Universe evolves towards complexity. There seems to be a direction in the evolution of the Cosmos but there is no *control* involved, only *attraction* to a purpose of a "higher" system.

It seems a higher purpose is always related to diversity, integrity, inter-connectedness, inter-dependence and coherence; or maybe the mystics are right and Everything is just One, but each time, on each iteration, it becomes a more complex One.

Information: The Key to Understanding Complexity

Intention

Intention is an attractor that can influence the system's performance. Intention is an *informational attractor* and, thus, affects energy and matter. *

Intention therefore can influence a system's performance in its material, energetic or informational state.

In our experience, whenever there is a leader or group of people that can influence the system from a higher perspective by rightly focusing on a common purpose there are more probabilities of creating a learning system and a new complex order. Conversely, whenever we lack this kind of leadership or if the leader does not have the right intention or mixes the purpose with her hidden agenda, it is very difficult to achieve the purpose.

Crime prevention, where things move quickly because the system iterates very frequently, is a great lab to test how intention works.

* *I had previously proposed that intention was an energy field, but I now find this new proposal more coherent with the proposal of Information, Energy and Matter.*

Consciousness

Consciousness is awareness of information. It is information seen or perceived by the receiver from a "higher" view. Consciousness is a highly evaded and mysterious subject.

David R. Hawkins, in his book *Power vs. Force*, makes a great attempt to structure consciousness in a fresh way, he proposes a *Map of Consciousness*. The map is an exponential scale from zero to 1000; zero having the more restricted or myopic vision, and 1000 being in resonance with the Universe's consciousness, the consciousness of the One.

Below 200, we resonate to the Ego; above 200, we resonate to the Self. It is also a threshold between lies and truth, force and power, control and order, death and life; he relates each level to emotions and to our view of life and God.

There are different "truths" at each level. For example, if our level is level 20 or *shame*, our emotion is *humiliation*, our life-view is *miserable*, our God-view is *despising* and the process is *elimination* ("shame kills you"). At 100 or *fear*, our emotion is *anxiety*, our life-view is *frightening*, our God-view is *punitive* and we *withdraw* from the world.

At 200 or *courage*, things change radically and our emotion is *affirmation*, our live-view is *feasible*, our God is *permitting* and we become *empowered*.

Acceptance, at 350, is a good level of consciousness since the average of humanity is circa 220 (it was below 200 before 1989). God becomes *merciful*, life becomes *harmonious*, we learn to *forgive* and our process is of *transcendence*.

Science is set between 400 and 499, the level of *reason*. God is *wise*, life is *meaningful* and we *understand* through *abstraction*. Science not only searches for truth, it accepts and invites counterarguments.

So what can be higher than science? *Love* at 500, *Joy* or *Unconditional Love* at 540, *Peace* at 600 and *Illumination* from 700 to 1000. In this last level, God becomes the *Self*, life just *is*, and we achieve *pure consciousness*.

I relate the map to purpose and intention. When we reach higher levels as individuals or in our work or creations, we are really aligning with a higher order. Our information then is much more powerful and has a greater impact on our environment.

The map is very practical and proposed to be used in daily life to respond to others, to create meaning in our life, or to design –as we have done- better organizations and programs.

Hawkins also shows us the way to calibrate through *kinesiology* and this is a mystery in itself if you try to understand it from a materialistic approach, but is quite reasonable from an informational perspective:

our body has access to the I-field; we are connected to Everything.

Information: The Key to Understanding Complexity

Truth

Truth is a strong attractor in the Universe. The Universe responds positively to truthful intention and information.

The human body responds positively to truthful environments and statements. So do all other social systems: relationships, families, companies, communities, nations and governments. The higher the level of consciousness, "truths" become more powerful.

In a social system, when something fails, when there is chaotic disruption or persistent failure, it usually has to do with untruthful intention or information, or both.

This seems to contradict the "useful lies" of social systems where individuals have to lie to blend in, groom others and survive. But while the Ego needs lies, the Self is always truthful and candid.

If the social system has a low level of consciousness it will require many lies to work, if on the contrary, its level of consciousness is high, there is acceptance, reason, love, peace and illumination.

Information: The Key to Understanding Complexity

Santiago Roel R

Is it a conscious Universe?

Why shouldn't it be? Why should intelligence, consciousness and knowledge be limited to a human perspective? Why should it be located in the brain or the nervous system? Who says so and who has proven it? No one. Why should it be determined or restricted to matter or energy? Why should it be located in matter and transported by energy? Why does it have to be even tied to "living" organisms, as we know them?

Hasn't quantum mechanics proven that we affect the behavior of a particle when we observe it? Hasn't it also proven that photons and even atoms become coherent in an informational way?

Intelligence, knowledge and consciousness all are related to information, and information is both independent from energy and matter. So, our view expands and becomes drastically different if we see information behind the manifest Universe: an atom is an atom because there is coherent information that in-forms that atom; a bird is a bird, because there is an informational field that forms that bird; the Sun is the Sun because there is an informational Sun. The Universe is an Informational Universe as it is also an energetic and material Universe.

If information is "within" or implicit in everything why can't it also be intelligent or conscious? It is not a mind-over-matter situation it is an information-over-information phenomena.

So maybe it is not a Big Bang but a Big Thought from a Conscious Universe.

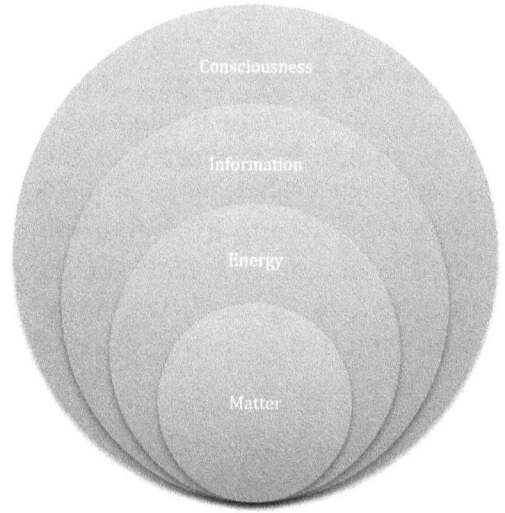

In this diagram, the higher dimension is consciousness, then information, energy and finally matter. In classical mechanics, causality is studied in the horizontal plane or the dimension of matter. In quantum mechanics, phenomena are *non-local, non-continuous* and *non-causal*; they seem to appear from the higher dimension of the plenum, the sea of energy. I suppose what I am proposing is the existence of two higher dimensions, two larger underlying seas: information and consciousness.

Santiago Roel R

In this way, we could observe and understand the Universe in its *vertical* causality the way it manifests or unfolds from consciousness, into information, into energy and finally, into matter.

How does all of this come together?

A review

Going back to our proposal: complex order is created by intention or purpose, rules of action and information. We have focused too much on matter and energy and overlooked information. Information is found within matter and energy.

The future, the past and the present create complex order. Rules of action are created by iteration. The purpose is projected into future iterations. Novel information is being created in the present. It is a bottom-up and top-down, and a focused and random process.

All three of these elements are really information. Information is everywhere and is the basis of our Universe. The Universe is entangled by information. The Universe learns by iterating. Laws are not universal or eternal they change through time and iteration. Information is also knowledge and consciousness and is not limited to matter or energy. There are levels of consciousness, the higher the level the more encompassing it becomes. The manifest Universe seems to evolve into a more complex, interconnected and conscious place. The

Universe is not controlled but is attracted to complex order.

What can we view from this perspective?

Science

Science could greatly benefit if it detaches itself from the strict energy-matter and the atomistic paradigms. Instead, we can start trying to make sense from a connected Universe and from the acceptance of underlying dimensions of consciousness and information.

It is time to move on and explore information not by linking it to entropy, or viewing it from Claude Shannon's transmission perspective, or Norbert Wiener's desire to control through feedback mechanisms, but from its capacity to create order in the Universe.

Complexity authors give credit to information but fall short of fully embracing what I propose: Information is key to understanding complexity, and probably, in helping us to understand the Universe. Most of these authors are also stuck in the bottom-up emergence paradigm and fail to recognize what seems more common, systems have a specific purpose and systems are part of a larger system; all of this is a top-down element to be acknowledged in emergence. I confirm my initial proposal: complexity is basically created by the interaction of rules of action, intention and information, and yes, all three of these elements are really the same: information.

Thanks to quantum mechanics we begin to understand that there is something deeper and more meaningful than matter or energy, and yet *normal* science is stuck in that bifurcation and afraid to make the leap. Non-conforming scientist or philosophers are ignored or ridiculed by the fearsome scientific community.

Religious science is based on belief not on reason; its intention is to re-link or tie down (*re-ligare*) people, not ideas. We don't have to embrace old religious myths or new scientific fundamentalisms to do this; we just have to open up to new proposals.

We could also start looking at vertical causality instead of excessively horizontal causality.

In this Promethean leap, I am sure many actual unsolved mysteries will be better explained.

Learning

We have all suffered from traditional education. We have all been bored, constrained, reduced, limited, and choked by education. Education has become a bureaucracy full of beliefs, hierarchies and outmoded industrial, mechanistic, paradigms.

We can create much better learning environments with intention, information and the free iteration of the system. We need creative individuals, not conformed erudite followers. Learning is natural, it does not have to be taught just enhanced and allowed.

Business and Government

Many large organizations have become truly chaotic and ineffective. There is so much to gain if we can learn to organize ourselves without the need for hierarchical control and deterministic plans. Organizations should enhance learning at all levels.

It is so much easier to work in a natural environment -more in communion with the way the Universe "organizes" itself.

Intention, information and free iterations are key to creating better organizations. Random and focused learning should be kept in mind to create intelligent and knowledgeable organizations.

Society: too much Yang

We have been living in the Yang or *masculine* principle: Top-down control, fixed and immutable laws, strategy, reductionism, separateness, single dimension, mechanical, fight-or-flight, domination, reaction. We have to balance this with a more feminine or Yin principle: self-organizing structures, flexibility, emergence, organic processes, holism, get-together-and-talk, multi-dimensions, prevention and integration.

Complexity is not and should not be about self-regulation, which is a masculine concept. It is about self-organization. It is not about strategy, but rather intention and emergence. It is not about detailed analysis and division into parts, but wholesome understanding.

Instead of focusing so much on rules and processes, we should enhance intention and information as key elements for order. I am not proposing a radical move from Yang to Yin, I am just saying we are unbalanced. This imbalance creates disintegration and discontent as we have been seeing in the world lately. Who could have predicted the latest social and political movements? And although some seem inarticulate in their demands and disorganized, there is a leitmotiv: the masses are discontent with hierarchical control, chaotic regulations, inequality by domination and insensitive government and corporations. We will see much more of this in the following years with two archetypes in play: on one hand, the Plutonic basic needs, the survival instincts, the animal chakras, and on the other, the Promethean revolutionary fire, the third eye chakra, the higher mind trying to liberate and empower mankind; the animal and the divine in tension, trying to resolve at the heart.

Information: The Key to Understanding Complexity

Afterword

As I finish this essay, I see several sea birds effortlessly coasting in the sky while the waves break rhythmically on the shore.

I can clearly see a purpose, basic rules and novel information in each bird as they hover gracefully over the setting sun. I can also see these principles in the lush tropical vegetation that surrounds me.

What about the ocean? Does it have a purpose or just rules of action and information in a random play? Or does it share a higher purpose, the purpose of life on Earth?

And what about the Sun and everything else that I can perceive with my senses?

How does information link all of this together? What about consciousness?

And then I stop making an effort and just sink into a meditative state. I stop thinking and begin to sense; all of my body becomes present. At that moment I stop being a separate entity and somehow merge harmoniously with my environment in a subtle and pleasant way. I can feel a tingle all over and my body seems to expand as it liberates itself from matter.

As I go deeper, duality ceases. There is no more inner and outer world. I am no longer separate and alone: I am the wave and the light and the flight, the specific and the abstract. Everything that was controversial is diluted and fades away in a totality. Time has no significance: there is no past, present or future. There are no more questions. There are no more words or thoughts or connections or dimensions. There is no manifest and potentiality.

Everything just is, and this feels much more *real*, it feels like *home*.

Santiago Roel R

Final Note

This essay was created with:

1. Intention: to understand the role of information in complexity

2. Rules of Action

-Truthful and honest to my own level of consciousness

-Easy to understand

-Based on experience

-Documented

-Calibration on Hawkins Map of Consciousness above 700*

3. Novel information

-No pre-determined index

-Writing was done each morning after meditation

-Open to intuition and synchronic events

4. Iteration

143 iterations from de first draft to the final version

Actual calibration of essay according to Hawkins Scale: 920

Information: The Key to Understanding Complexity

Bibliography

1. Axelrod, Robert & Cohen, Michael D. *Harnessing Complexity*, Basic Books, 2000.

2. Abraham, Ralph. *Chaos, Gaia, Eros*, Harper San Francisco, 1994.

3. Bar-Yam, Yaneer, *Making Things Work*, Knowledge Press, 2004.

4. Bohm, David. *Wholeness and the Implicate Order*, Routledge Classics, 2002.

5. Briggs, John & Peat, David. *Seven Life Lessons of Chaos*, Harper Collins e-books, 1999.

6. Burger, Edward B. & Starbird, Michael. *Coincidences, Chaos and all that Math Jazz*, Norton, 2006.

7. Chaitin, Gregory. *Meta Math, The Quest for Omega*, New York: Vintage Books, 2005.

8. Fields, R. Douglas. *The Other Brain*. Simon & Schuster, 2009.

9. Ford, Kenneth W. *The Quantum World: Quantum Physics for Everyone*, Harvard University Press, 2004.

10. Gladwell, Malcolm. *The Tipping Point*, New York: Bay Back Books / Little, Brown and Company, 2000.

11. Gleick, James. *Chaos: Making a New Science*. New York: Penguin Books, 2008.

12. Gleick, James. *The Information: A history, a theory, a flood*. New York: Pantheon Books, 2011.

13. Gribbin, John. *Deep Simplicity: Bringing Order to Chaos and Complexity*, New York: Random House, 2004.

14. Haisch, Bernard. *The God Theory*, Weiser Books, 2006.

15. Hawkins, David R. *Discovering the Presence of God*, Veritas Publishing, 2006.

16. Hawkins, David R. *Healing and Recovery*, Veritas Publishing, 2009.

17. Hawkins, David R. *Power vs. Force*, Hay House Inc, 2002.

18. Hawkins, David R. *Reality, Spirituality and Modern Man*, Axial Publishing Company, 2008.

19. Hawkins, David R. *Truth vs. Falsehood*, Axial Publishing Company, 2005.

20. Hellinger, Bert. *La Paz Inicia en el Alma*. Mexico: Herder, 2006.

21. Hoover, Thomas. *The Zen Experience*, The New American Library, 1980.

22. Jung, Carl G. *Modern Man in Search of a Soul*. Harcourt Brace Jovanovich, Publishers, 1993.

23. Jung, Carl G. *The Archetypes and the Collective Unconscious*, Princeton University Press, 1990.

24. Kauffman, Stuart. *Reinventing the Sacred*. Basic Books, 2008.

25. Kiel, D. & Elliot, E. *Chaos Theory in the Social Science*, The University of Michigan Press, 2007.

26. Santa Fe Institute. Editors: Langton, Christopher G; Taylor, Charles; Farmer, J. Doyne; Rasmussen, Steen. *Artificial Life II*, Addison-Wesley Publishing Company, 1992

27. Khun, Thomas S. *The Structure of Scientific Revolutions*, University of Chicago Press, 1970.

28. Lazlo, Ervin. *Quantum Shift in the Global Brain*, Inner Traditions, 2008.

29. Lazlo, Ervin. *Science and the Akashic Field*, Inner Traditions, 2007.

30. Lazlo, Ervin. *The Akashic Experience*, Inner Traditions, 2009.

31. Lipton, Bruce H. *Spontaneous Evolution*, Hay House Inc, 2009.

32. Lipton, Bruce H. *The Biology of Belief*, Hay House Inc, 2008.

33. Lorenz, Edward N. *The Essence of Chaos*, The University of Washington Press, 1995.

34. Mc Taggart, Lynne. *The Field*, Harper-Collins Publishers, 2008.

35. Mc Taggart, Lynne. *The Intention Experiment*, Free Press. 2007.

36. Mandelbrot, Benoit & Hudson, Richard L. *The (Mis) Behavior of Markets, A Fractal view of Financial Turbulence*, New York: Basic Books, 2004.

37. Mitchell, Melanie. *Complexity: A Guided Tour*, New York: Oxford University Press, 2009.

38. Payne, John L. *Constelaciones Familiares para Personas, Familias y Naciones*, Spain: Ediciones Obelisco, S.L., 2009.

39. Prigogine, Ilya. *The End of Certainty: Time, Chaos, and the New Laws of Nature.* New York: The Free Press, 1997.

40. Radin, Dean I. *The Conscious Universe*, Harper Collins, 2009.

41. Rosenblum, Bruce & Kuttner, Fred. *Quantum Enigma*, New York: Oxford Univesity Press, 2006.

42. Roel, Santiago. *Between Order and Chaos: A Mexican crime-prevention success story.* www.prominix.com, 2008.

43. Roel, Santiago. *Can we Change Social Systems?* www.prominix.com, 2010.

44. Seife, Charles. *Decoding the Universe*, USA: Penguin Books, 2006.

45. Sheldrake, Rupert. *The Presence of the Past*, Park Street Press, 1995.

46. Sheldrake, Rupert. *Morphic Resonance*, Park Street Press, 2009.

47. Sheldrake, Rupert; McKeena, Terence; Abraham, Ralph. *The Evolutionary Mind*, Monkfish Book Publishing Company, 2005.

48. Schrödinger, Erwin. *What is Life?* Cambridge University Press, 1967.

49. Stein, Murray. *Jung's Map of the Soul*, Carus Publishing Company, 2010.

50. Stevens, Anthony. *Jung: A very Short Introduction*, Oxford University Press, 1994.

51. Strogaz, Steven. *Sync: How Order Emerges from Chaos in the Universe, Nature and Daily Life,* New York: Hyperion, 2003.

52. Taylor, Marc C. *The Moment of Complexity.* London: The University of Chicago Press, 2003.

53. Tarnas, Richard, *Cosmos and Psyche,* USA: Viking Penguin, 2006.

54. Tzu, Lao. *Tao Te Ching,* Ego Books, 2009.

55. Volkenstein, Mikhail V. *Entropy and Information,* Switzerland: Birkhauser Verlag AG, 2009.

56. Von Bertalanffy, Ludwig. *General System Theory,* New York: George Braziller, 1969.

57. Waldrop, Mitchell. *Complexity: The Emerging Science at the Edge of Chaos.* New York: Touchstone, 1992.

58. Walker, Evan Harris. *The Physics of Consciousness.* New York: Basic Books, 2000.

59. Wiener, Norbert. *Cybernetics or Control and Communication in the Animal and the Machine,* 2nd edition, MIT Press, 1961.

60. Wilber, Ken. *A Brief History of Everything,* Shambhala Publications, Inc., 2000.

Santiago Roel R

61. Wilber, Ken. *The Essential Ken Wilber: An Introductory Reader*, Shambhala Publications, Inc., 1998.